Práxedes Maribel Mora Plúas
Leonardo Conde Suárez
Jacqueline Suárez Herrera

Incidence of Technology as a Pedagogical Tool

Práxedes Maribel Mora Plúas
Leonardo Conde Suárez
Jacqueline Suárez Herrera

Incidence of Technology as a Pedagogical Tool

to Facilitate Mathematics Learning

ScienciaScripts

Imprint
Any brand names and product names mentioned in this book are subject to trademark, brand or patent protection and are trademarks or registered trademarks of their respective holders. The use of brand names, product names, common names, trade names, product descriptions etc. even without a particular marking in this work is in no way to be construed to mean that such names may be regarded as unrestricted in respect of trademark and brand protection legislation and could thus be used by anyone.

Cover image: www.ingimage.com

This book is a translation from the original published under ISBN 978-613-9-40116-1.

Publisher:
Sciencia Scripts
is a trademark of
Dodo Books Indian Ocean Ltd. and OmniScriptum S.R.L publishing group

120 High Road, East Finchley, London, N2 9ED, United Kingdom
Str. Armeneasca 28/1, office 1, Chisinau MD-2012, Republic of Moldova, Europe
Printed at: see last page
ISBN: 978-620-7-70079-0

AUTHORS

PRAXEDES MARIBEL MORA PLUAS

MAIL: MARYMORAPLUAS73@GMAIL.COM HTTPS://ORCID.ORG/0000-0003-1758-819X EDUCATIONAL UNIT "TENIENTE HUGO ORTIZ

COUNTRY ECUADOR

JACQUELINE VIKY SUAREZ HERRERA

E-MAIL: JACQUELINESUAREZHERRERA@GMAIL.COM

HTTPS://ORCID.ORG/0009-0000-4614-4411

UNIDAD EDUCATIVA DEL MILENIO "RAÚL ALFREDO VERA VERA"

COUNTRY ECUADOR

LEONARDO OMAR CONDE SUAREZ

MAIL: LEONARDOCONDESUAREZ@GMAIL.COM

HTTPS://ORCID.ORG/0000-0002-0693-7764 UNIDAD EDUCATIVA

FISCAL "LOS VERGELES".COUNTRY ECUADOR

CONTENT

SUMMARY

The importance of technology as a pedagogical tool to facilitate the learning of mathematics through the design of strategies and implementation of seminars and workshops for teachers and legal representatives. The purpose is to incorporate ICT tools (Information and Communication Technologies) in the work of the teacher in order to provide students with the knowledge, skills and procedures necessary to develop them in the didactic guide of the educational institution. The contents of the workshops are broken down in a sequential and schematic way, in which the content of a didactic guide will be socialised in which the use of tools available on the web for the teaching-learning process of mathematics is made known. Each of these tools will allow students to motivate and innovate educational teaching in the new millennium and at the same time take advantage of these potentialities, allowing them to promote collaborative work in the network for the exchange of ideas and experiences through the appropriate use of technology in the subject of mathematics. Educational technology is designed to help educators to plan and control the learning process in a more efficient way, being this possible thanks to the use of resources such as smartphones, computers, televisions, among others. The development of these techniques has been established for decades to optimise the presentation and understanding of educational content to students.

Keywords: use of technology; strategies; learning; pedagogical tools; workshop seminar

INTRODUCTION

Trends in teaching are currently oriented towards strengthening competencies, knowledge and fundamental values for learning. Technological advances are identified as a valuable resource capable of accompanying the teaching of different areas of study at any educational level, which unquestionably calls for a revolution both in research and in teaching, in order to take advantage of the potential offered by technologies as pedagogical tools in the educational field.

The appropriate use of technology in the mathematics classroom depends on the teacher, with any te3 and web applications to design tasks in an interactive way for students.Students using technological tools appear to work independently of the teacher; this is a misleading impression, the teacher plays important roles in a classroom that is enriched with the use of technology.technology, making decisions that disrupt the learning process of students in important ways is the teacher who decides if, when or how to use technology; students use calculators and computers in the classroom, the teacher assumes the opportunity to observe how they perform their mathematical logical reasoning for some is difficult to observe in other circumstances where the same expected results are not obtained.The technology allows teachers to examine the methods students have followed in their mathematical investigations, as well as the results achieved, thus enhancing the information available in this area of the curriculum that they use when making teaching decisions.

TECHNOLOGY

The digital age has changed all aspects of our daily lives and education is no exception, as the industrial part is being replaced by the computer age, education is facing great challenges and needs; being part of the digital innovation by embedding technology, known as educational technology in education. The integration of new technologies in schools has changed teaching methods to such an extent that it has made room for a digital culture in schools.In the classroom, such an explanation is not sufficient to understand their influence today, so we need to discover their role today. Solutions to educational problems are located in the use of information technology, i.e. the use of computers or various telecommunication devices to store, transfer and manipulate data, for educational purposes.Currently, staff in different educational centres have access to the Internet, computers, digital whiteboards, mobile phones and tablets to share knowledge or organise lessons and tasks; successfully adapting educational methods to the digital age, providing teachers and students with learning resources where they can obtain information on the introduction of new technologies, which also open up new spaces for recreation and expression, such as games and blogs in their teaching and learning.Technology responds to the desire and will of people to transform our environment, transform the world around them, looking for new and better ways to satisfy the desires, the different needs of development, design and implementation of products obtained through the application of strategies, methods and processes used for their easy application.According to (Ruiz, 2012), in these times when our governments are trying to dismantle the technological teachings, it is not superfluous to explain that it is the Technology, why it is important, and why it should be taught. The authors emphasise that technology is a set of knowledge and skills that, when used logically and methodically, enables human beings to change their environment.Technology is human activity and its results, so there are many systems that allow us to interact, move, dress, eat and create new things according to the needs of today's society; it is one of the most powerful, versatile and vital resources of our species, capable of transforming our environment and even our own body and mind. It is the result of a long-term cultural and scientific development and represents an enormous power and risk.Technology has also become an object of everyday consumption these days triggering the emergence of a technology market and a consumer culture of so-called gadgets and devices, more or less for decorative and recreational purposes, but cutting-edge technology continues to strive to fulfil humanity's long-delayed dream of curing diseases, improving the quality of life and exploring the frontiers of space, and in the process, it is harnessing exponential scientific knowledge.From this perspective (Maricel Occelli & Gatica, 2019) mentions that: The educational event is an essentially communicational phenomenon, which, as such, is The ICT-mediated environments that we learn to use in the framework of social interactions made possible by communication. These interactions take place in a dialogical and participatory process within the virtual context. Social networks are an ideal space to visualise these relationships and occupy a prominent place in personal life. We are aware of the amount of time our students spend on social communications, the power of the networks and the information that is established in them. Our interest in these aspects relates to the study of social networks as teaching and learning tools.We began exploratory work by taking five Facebook teaching groups and analysing the types of communication, interaction and information presented in them (Movsesian and Valeiras, 2015).Technology marked a before and after the global pandemic, the same that only served for social networks, business and personal mail, social networks had as a priority to publicise the movements, emotional states, luxuries of visits to its users; remembering that most of the teachers did not know the basics of computer science, which led them to re-educate themselves in this technological area. The radical change brought about by the social networks that served as an information bridge between teachers, students and parents, different entities The public universities created technological courses and the management of different educational platforms; nowadays, office automation tools are applied in the classrooms in order to provide students with new knowledge.

Principle of Technology

The use of computers increases the social disparity between those who have access to this technology and those who do not. Today, this disparity is no longer limited to those who own computers and those who do not, but also to those who own multimedia computers connected to the Internet. Research in mathematics shows that the new computer programmes are didactic agents that create new situations that cannot be achieved with traditional means such as pencil and paper, which makes the child a passive entity. As a result, it is believed that educational computing can significantly help improve teaching and learning by becoming essential to the teaching and learning of mathematics because it enhances student learning. Computers, calculators and other electronic technologies are essential tools for teaching, learning and "doing" mathematics; they provide visual images of mathematical concepts, facilitate the organisation and analysis of data and enable efficient and accurate calculations. Mathematics, including numbers, measurement, geometry, statistics and algebra, are fields that can help students focus when they have technological tools. Mathematics teaching applications should be used frequently and responsibly in order to enhance students' learning of mathematics.Technology exists, it is versatile and powerful, making it possible and necessary to re-examine what mathematics students should learn and how best to learn it. Every student in mathematics classrooms based on the principles and standards has access to technology to facilitate their mathematics learning, guided by an experienced mathematics teacher.

Importance of technology.

Nowadays, technology is one of the most important factors for the institutional use of institutions, but is it really managed properly in today's society? Many people who can use technology only think about the internet, but never think about how important technological tools are for the management of an institution's information and its social projection. Technology has had a significant impact on human life in various ways, leading to significant change, overcoming some obstacles that affect reality and demonstrating impressive advances for society in general. Today, technology can be used in everything that is done, helping and facilitating the day to day, developing new capabilities and providing solutions to the institutions, entities or beneficiaries that use it. The value of technology is almost always related to its useful uses; in fact, the complicated implementation of expensive or complex technologies makes it difficult to succeed, so that sometimes seemingly rudimentary technologies prevail over much more "modern" ones. In any case, technology is often in constant evolution, which refers to the practical development of new ideas created by different scientific disciplines. It is therefore closely related to the concept of technological innovation in education.

Pillars of educational technology

Technologies have changed the field of education the integration of educational technologies in schools and their contribution to the education of children and young people. education; it is not without controversy; there must be a consensus on the need to of using technological advances for learning in the classroom.Educational problems are solved by using information technologies, i.e. the use of computers and other telecommunication equipment for data storage, transmission and manipulation, we understand that the use of technological devices is for educational purposes.The new educational model has created three technological areas such as programming, robotics and 3D printing, these areas being the fundamental pillars of educational technology.

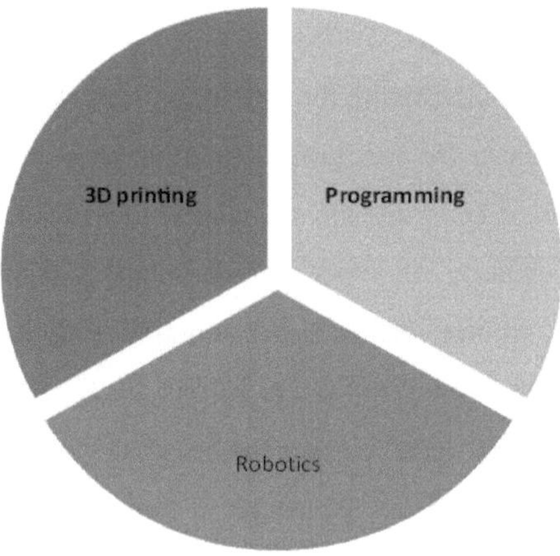

The educational model takes into account three highly related areas: technology, programming, robotics and the internet of things, being predictable subjects and aiming to develop young people's skills to succeed in the information society, these areas are the pillars of educational technology:

PROGRAMMING

Programming is the creation of a program or application by developing source code based on a set of instructions that a computer executes to run the program; it allows the computer to run and perform tasks requested by the user. From the position of (VIVES, 2021) he states the following: One of the characteristics of robotics is that it is taught through gamification, i.e. learning is done through play. This makes it possible to assimilate mathematical, physical, mechanical or computer concepts in a fun way and thus improve the acquisition of skills that form part of school curricula. From the point of view of (Education, 2020) it states that: Most productive activities in the future will require basic coding and programming skills. In addition, many of the jobs that will be performed in the future will require basic coding and programming skills. The younger generations do not yet exist, which is why programming is part of the skills of the future, where computer knowledge and competences will be indispensable. By teaching programming is to prepare students for the technical world of work, they learn to identify errors in complex problems, process the process of self-correction, the function of programming is to promote the learning of logic, creativity, solution finding and entrepreneurship.The act of scheduling, that is organising a sequence of actions to achieve something, can be used in many contexts, typically when organising an excursion, a holiday, talking about a programme or a list of programmes on a television channel and their broadcasting schedules, or a list of films in a cinema; in computers, scheduling is also an important part of the computer-user relationship.

Robotics

Robotics is a discipline that teaches a set of instructions to autonomously run a device or robot, in addition to learning to use the correct language, programming robots also allows students to visually see programming errors and their limitations; the digital era requires people who know how to programme these devices to provide solutions to the growing science and engineering demands of the future workplace.From the position of (VIVES, 2021) he states the following: One of the characteristics of robotics is that it is taught through gamification, i.e. learning is done through play. This makes it possible to assimilate mathematical, physical, mechanical or computer concepts in a fun way and thus improve the acquisition of skills that form part of school curricula. Robotics is an exciting field that combines engineering, computer science and creative design, focusing on the creation, programming and operation of robots, which are machines that can perform tasks autonomously or semi-autonomously, some of the aspects of robotics.

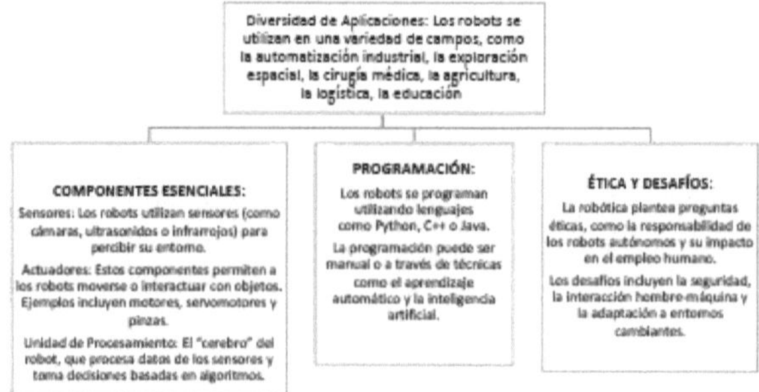

Educational robotics is the use of robots in the classroom with the aim that students learn through play and acquire knowledge related to mathematics, technology, science and engineering, becoming an interdisciplinary subject that encompasses both mathematical and scientific concepts as well as cognitive skills.

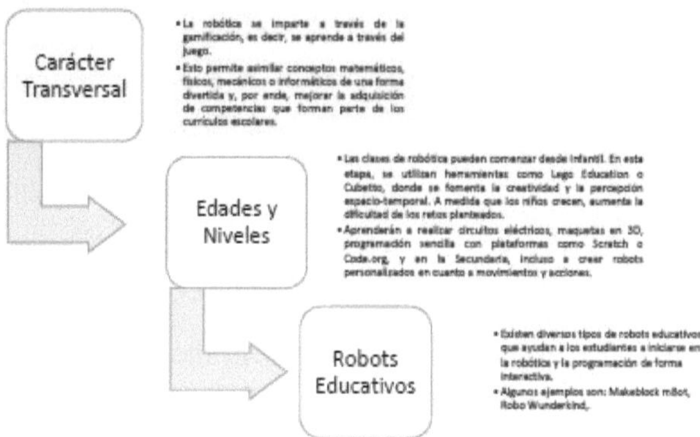

In short, educational robotics not only teaches technical concepts, but also skills such as creativity, logical thinking and problem solving, an effective tool to motivate students and prepare them for an increasingly technological world.

3D printing

3D printing is a set of additive manufacturing technologies that create three-dimensional objects by stacking successive layers of material, in the process of creating physical objects by layering materials in a digital model, thus the process of creating a physical object in three dimensions from a digital object or model using a 3D printer, which can use different techniques and material overlays to create a perfect replica.From the point of view of (Berchon, 2014) he states the following: 3D printing is a set of additive manufacturing technologies that create three-dimensional objects by stacking successive layers of material. The process of creating physical objects by layering materials in a digital model. It is therefore the process of creating a physical object in three dimensions from a digital object or model using a 3D printer, which can use different techniques and layering of materials until a perfect replica is created. The range of applications of 3D printers in industry, medicine and even everyday life is so wide that some experts dare to claim that the use of this technology can revolutionise industry and the economy, as it can reduce production costs and influence production; get jobs. In countries that produce goods cheaply, on the other hand, there is also a debate about how dangerous this tool is - anyone can build a weapon using blueprints found online.In the 21st century, education must transfer increasing knowledge on a large scale and in an efficient way by developing theoretical and technical knowledge adapted to cognitive societies, building a basis for future skills and at the same time, it will be necessary to find and define guidelines to set the parameters at a global level. For Unesco in 1996 established a solid foundation in 21st century education such as the pillars of education:

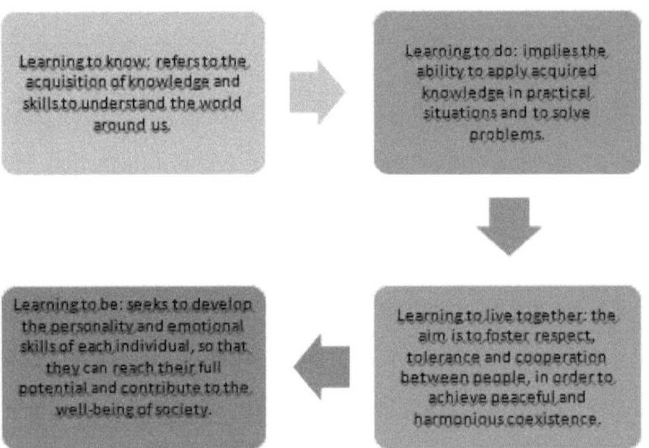

COMPONENTS OF EDUCATIONAL TECHNOLOGY

Currently, technologies for learning and knowledge are taking centre stage and provide daily sustenance in the fields of culture, society, sport, entertainment and, of course, information; however, at the educational level, concepts such as student or teacher have changed, becoming learning guides, reinforcing and practising concepts such as cooperative work. Educational technology does not start only with the use of computers, but everything related to educational technology allowing the clearer development of classroom activities, improving the teaching and learning process; today educational technology is used in pedagogy because it helps to improve learning and the quality of education, due to the fact that it gives the young people the facility to understand the subject with the support of audio, video and images.

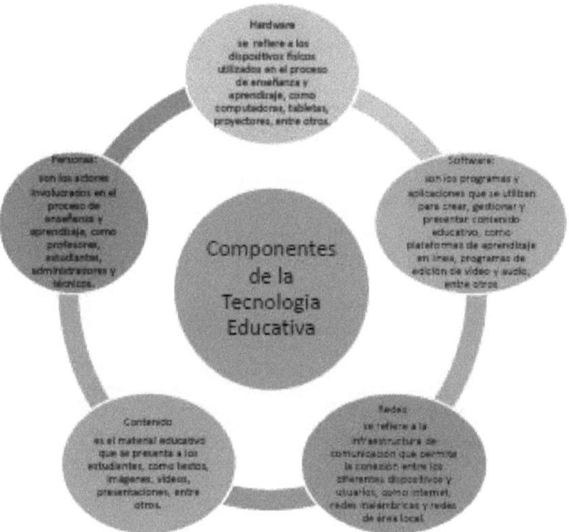

Educational technology helps with the transformation of teaching and learning, allowing for better personalisation and adaptation to the individual needs of learners, however, remember that technology does not fully replace human interaction, its use must be carefully planned and evaluated to ensure its effectiveness.

Types of educational technology

Technology is increasingly being used in all aspects of life, including education, educational technologies enhance the learning process by offering new ways of interaction, technological tools and more effective teaching methods, online education is becoming an increasingly popular option for those seeking a more flexible and affordable education. The age of technology has made it easier to pursue online education in a variety of ways, however, it is worth noting that there are few institutions specialised and committed to providing students with the best methods of academic training. With this in mind, it should be added that there are many types of educational technology available to us, the most notable of which are:

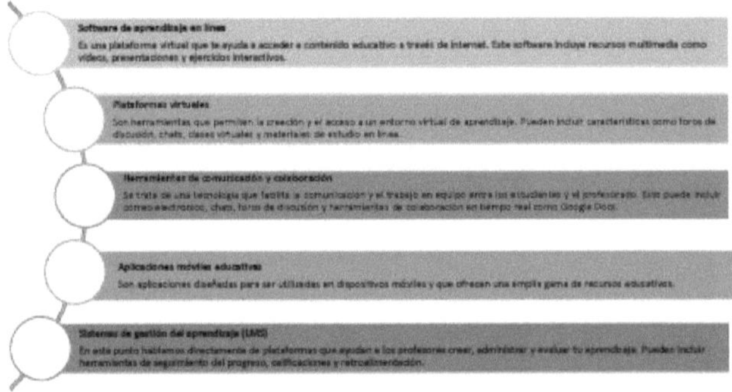

According to the document (Chavero-Tapia, 2020) describes that: A new, broader conception of education should lead each person to discover, awaken and increase their creative possibilities, thus actualising the hidden treasure in each of us, which implies transcending a purely instrumental vision of education, perceived as the obligatory way to obtain certain results (practical experience, acquisition of diverse skills, etc.), economic purposes), in order to consider its function in all its fullness, namely the realisation of the person who, as a whole, learns to be (Delors, 1996, p. 92)The use of technology can enable them to learn about other cultures, learn stories of people who have changed the world in some way, inspire and change people's attitudes during the learning phase and help them to better create their vision as professionals; it requires continuous training, and the sheer volume of information processed daily by all branches of science underlines its paramount importance as an educational resource one has to look at the evolution of companies to understand and admit that obsolete training can no longer be continued. From the age of craftsmanship to our own, where globalisation dominates the environment, ideas, technological innovations have taken hold and it is impossible to exclude them from the educational stages that every future qualified professional must go through; Conceptualisation of ICTs (Information and Communication Technologies) Today's society is characterised by the use of new information and communication technologies (ICT), which requires from its citizens a series of personal, social and professional competences to cope with the changes imposed in the fields of science and economy. Following this ever-changing world, there is an education system that prioritises teachers, who are seen as the engine of educational change.Therefore, it is imperative that educational institutions do not fail to take advantage of technological advances, teachers are gradually beginning to use new technologies to promote process development. Everyone can access information, teachers require constant professional growth the 21st century revolution will be to understand diverse learning styles, design and organise lessons, set goals and achieve them, motivate students and encourage participation, develop student-based learning, help students use different resources and sources, encourage self-learning, self-assessment and research both in the classroom and outside the classroom.Educational researchers can create this research profile using a variety of approaches e.g. classroom activities, journals, questionnaires, experiments, individual studies and new technologies are methods, it is important to note that the focus of this research is on education that implements strategic changes in didactic strategies so that communication and distribution of materials (texts) emphasise the availability and potentials of technology in the different subjects of the curriculum. The use of ICT in subjects that have an impact on students' academic performance can be transformed from a simple text to mathematics teaching guide that encourages students to learn and access text-based communication, written communication, data processing, numeracy and numerical data analysis, statistical data analysis and graphical expression. These resources can be used the more innovation and technology are encouraged, the more attention and learning power students will have by using ICT in subjects that are often complicated and this will help students to learn and teachers to teach more effectively.

ICT IN EDUCATION

ICT in education is based on three raisons d'être in education:

• It is used for the digital literacy of students, who must acquire basic skills in the appropriate use of ICT (Information and Communication Technologies).

• It is productive as it takes advantage of the benefits they provide when carrying out activities such as preparing notes, exercises, (via e-mail) disseminating information (web teachers and students).

• Innovate in teaching practices, and the new didactic possibilities provided by ICT to get students to do better work and reduce school failure.

Educational use of ICTs in the teaching-learning process

From the beginning of writing to the present day, the process of creating artefacts, in the broadest sense of the term, that promote the preservation and circulation of information so that we can transform it into useful knowledge has been a constant activity, some believe that computer and communication technologies (ICTs) are not so new. According to (Rossana, 2021)mentions that: In today's society, the alliance between technology, information and knowledge has been given in a fundamental way for the work and personal environment of any professional, however, every day there are technological innovations that demand the constant updating of knowledge. Therefore, the educational environment must be updated in terms of the changes and advantages offered by ICT, being an important didactic and pedagogical tool, which is the reason for the development of this topic.Nowadays, all aspects of life require efficient management in order to communicate both nationally and internationally, and everyone can easily access global information through their computers, thanks to the almost universal availability of these resources, which allows us to keep up to date, however, in many less developed countries it is difficult to access this information. There is a reality that is a common factor in all countries that takes place in rural areas due to their low levels of development in terms of technology, the lack of technological equipment and Internet access lines, the scarce training of human resources and the few contents referenced to t h e realities of rural territories, thus increasing a large gap between the urban and rural sectors.Information and communication technologies (ICT) are becoming a vital factor in the evolution of the new global economy with the rapid changes that are taking place in society; in the last decade, new information and communication technology tools have brought about a profound change in the way individuals communicate and interact in the business world, causing significant changes in different fields. As Alícia Martí (Alícia Martí, 2020) says about: the use we make of technology: whether we use it exclusively to present content or whether we use it so that students can access information critically and even create and share documents. "It is necessary to bear in mind that in many cases ICT is used to support lessons and is integrated into a traditional teaching model. But technologies should become a catalyst for pedagogical innovation and be used in collaborative projects.Although ICTs are of great social and economic importance, they are not adequately and satisfactorily used in the United States and many other countries around the world. Computer technology is part of the environment in which life takes place, so there is an increasing urgency to learn to live with it and to exploit its undoubted potential.It is also important to note that learning to teach is widely believed to be a lifelong experience. That is why teachers should experiment in the classroom, as mentioned by (NUEZ, 2008): The design of multimedia applications presupposes the conjunction of two inseparable parts: pedagogy and technology. Pedagogy takes into account the selection of educational methods used to achieve the participation of the student as an active entity and the use of technology implies the use and combination of the modalities of educational informatics to achieve an informatics product that meets the proposed objectives.New computer technologies allow the construction of shared virtual environments where the concept of a "learning environment" can be applied. Time limits and remote access are virtually eliminated, allowing persistent communication across borders. In Ecuador, people are looking for ways to use technology, even if they do not have it directly, in homes they use ICTs in the same way they access electricity and water and have the opportunity to adapt to the use of new technologies. The INEC report found a gap in ICT access between urban and rural areas. Ecuadorians go online, check their email, read newspapers and magazines, look at movie posters and calendars, download music, browse photo sites, access Facebook accounts and chat with friends; education and learning are the main reasons people use ICTs (Information and Communication Technologies) at the city level.The foray into technology in the classroom is explained by the fact that schools and universities encourage students to make applications in the classroom. online, usage varies according to their age, they access Facebook accounts and chat with friends. Education and learning are the main reasons people use ICT (Information and Communication Technologies) at city level; this result can be explained by the fact that while schools and universities encourage students to apply online, usage varies according to age.Although access to ICTs is quite advanced in Ecuador, the lack of adequate teacher training to modernise teaching methods and technical equipment has hindered the introduction of educational ICTs at the level of basic education institutions in the public sector, with a very different reality in private and academic institutions where the use of ICTs predominates.

Importance of ICT (Information and Communication Technology) in the teaching and learning process.

Nowadays, information and communication technologies have become indispensable tools in educational institutions, as this resource opens up new opportunities for teachers to exchange ideas, methods, using ICT as a tool to favour the taking of decisions and the use of information and communication technologies in teaching and learning. decisions in the face of the educational needs of today's world in line with today's society.As expressed by (Ana VIÑALS BLANCO, 2016), on the teacher in the digital era she states that: We are facing a time of innovation in the fundamental pillars of the current education system. A change that must take into account not only the features of a networked society and the intrinsic characteristics of digital natives, but must also consider the demands of the labour market. Ultimately, teachers' aim is to prepare students for life, a digital life. So far, this adaptation has materialised in the creation of new basic competences linked, logically, to ICT and the notion of lifelong learning; competences which are now in force and which have replaced the old objectives as assessment indicators.This means that the scientific progress that is occurring in society entails radical changes in educational institutions, starting from the infrastructure for the implementation of new technologies and the most important one is the application of techniques and methods in the way of teaching and learning to make significant use of didactic resources in the different subjects contained in the curriculum. Therefore, new information and communication technologies will undoubtedly greatly improve the teaching and learning process in various educational institutions, because teachers use these innovations to stimulate students' interest in research and motivate students through interaction with the computer as participants in their own learning; engaging with the materials for teaching in the different areas of the curriculum. Considering that (Introduction: Children in the digital age, 2017) mentions that: 'Even as ICTs have fostered knowledge sharing and collaboration, they have also facilitated the production, distribution and sharing of sexually explicit material and other illegal content that is used to exploit and abuse children. Such technology has opened up new avenues for child trafficking and new means of concealing such transactions from law enforcement. It has also made it much easier for children to access inappropriate and potentially harmful content and, more surprisingly, to produce such content themselves. The use of technologies makes the educational process interesting for teachers because it guides the process that children discover through the use of ICT in the classroom, through constant research, students have access to a In this sense, the use of ICT prioritises the relevance of competences for meaningful learning, a challenge posed by technology itself.

ICTs according to curriculum updating and strengthening

Information and communication technology is a set of advances that includes new computer tools for communication, in this case education, and helps to develop macro skills in students. The idea is that there is diversity, but there is purpose that we live in a third millennium society characterised by accelerated change in the scientific and technological fields. According to the point of view of (Eduardo Parra Zambrano, 2015) indicates the following: In order to understand the process of curriculum integration it is first necessary to establish its conceptual bases. Sánchez (2002) defines the curricular integration of ICT as the process of making them an integral part of the curriculum, as part of a whole, permeating them with the educational principles and didactics that make up the learning process. This fundamentally implies a harmonious and functional use for a specific learning purpose in a specific domain or subject area. a curricular discipline. This definition emerges as a synthesis of the approaches to the concept previously established by the following authors: Grabe & Grabe (1996) state that integration occurs "when ICT sits comfortably with the teacher's instructional plans and represents an extension rather than an alternative or addition to them". For Merrill, P., K. Hammons, B. Vincent, P. Reynolds, P., L. Cristiansen, and M. Tolman (1996) this integration implies a combination of ICT with traditional teaching procedures to produce learning, attitude more than anything else, willingness to combine technology and teaching in a productive experience that moves the learner to a new understanding.The knowledge, tools and methods for performing and communicating mathematics are constantly evolving. For this reason, both the learning and teaching of mathematics focus on developing skills with the success criteria necessary for students to solve everyday problems while improving their logical and critical thinking skills.Knowledge of mathematics is not only satisfying, but also very necessary to be able to interact fluently and effectively in the mathematical world. Most of our daily activities require decision-making based on this science by creating a chain of logical considerations.Choosing the best options, buying products, understanding statistical and informative graphs in newspapers, determining the best investment opportunities, interpreting the environment, goods and works of art, i.e. applying them to the development of everyday life. Students have the right to receive the best mathematics education to enable them to fulfil their personal and professional aspirations in today's knowledge society, after acquiring competences that will facilitate access to a wide range of careers and professions. Therefore, all stakeholders involved in education, including authorities, parents, students and teachers, must work together to create appropriate learning environments. These rooms enable students of all abilities to understand and learn important mathematical concepts in collaboration with teachers qualified in their subject, they must teach and learn mathematics is a challenge for both teachers and students. Based on the principle of fairness, the use of technology and ICT in mathematics education is encouraged as it is a beneficial tool for both teachers and students.This tool can improve the process of abstracting, transforming and demonstration of some mathematical concepts, the area of mathematics is generally thought of as a subject that uses only blackboards and liquid chalk, but on the contrary, it requires the teacher to act as an intermediary, the facilitator to practice the use of educational ICTs and the slides to be reliably made. Easy-to-read text helps students visualise graphics and eye-catching colours that capture girls' attention, while technology is essential for meaningful learning. According to this definition of the Ministry of Education, the important thing is that education in general will bring about a significant change, where students are no longer included in the statistics of computer illiteracy, and technical resources are intended to be used in such a way that illiteracy is improved. By being creators of knowledge themselves, they expand their capacity, develop the skills and competencies to be critical citizens, and achieve maturity of knowledge in the various subjects included in the curriculum.

PEDAGOGICAL TOOLS

The most widely used tool is e-mail, which is very easy, fast and fluid to use, asynchronous; forums and chats allow real-time communication between many users, through which documents are sent, files are transmitted or images and sounds are attached; browsing through web browsers. They allow to focus on topics, in electronic publications, digital magazines, newsletters, distribution (or discussion) lists, databases and virtual libraries that are available on the network and can be accessed.As stated by (Vega, 2018), the important thing is that each pedagogical tool is created with the children's development process in mind, with the skills to be enhanced, with the experiences that the children live with their teachers in the classroom, in the open spaces of the centre and with their families at home. Pedagogical tools are a fun way to promote play. They offer open opportunities to build towers, to symbolise by drawing, personifying or making three-dimensional productions; to play with light and shadow, to imitate and develop motor games or manual precision.The teacher can use the internet within the classroom: developing presentations and interactive resources, making complex documents with calculations, maps or simulations, teaching new knowledge and reinforcing the The aim of the project is to promote the use of the teaching-pupil work, the creation of a web page or the use of office applications, promoting pedagogical innovation or the use of interactive whiteboards in educational institutions.One of the benefits that the Internet offers to education is the ability to collaborate (intranets), its usability greatly facilitates administrative tasks. Considering this situation, when I think about it carefully, I believe that the impact of changes in education is not only delaying investment in equipment and training, but also changes in attitudes and attitudes, and that process will take time, it is possible.

21st century pedagogical tool.

It is important to point out that some teachers confuse the pedagogical with the technological, given that they talk about incorporating new media such as computers and other support systems to generate an educational change, such change is a function of the pedagogical tool; in this proposal it is the Learning Investigation (AI) and technology is only the means to guide this educational process.Learning Research is a strategic plan to generate a dynamic of intellectual work on the problems of practice. professional. It is based on overcoming cognitive obstacles, professional skills and student learning in the exercise of scientific research in all areas.New tools and technologies facilitate learning as surveys and applications can be used to diagnose students' understanding in real time, which is key in e-learning, where much of the process is done on an individual basis.Using the words of (Seminarium, 2018) indicates that: On the other hand, there are virtual boards, such as Padlet, as well as shared documents, such as Google Docs, which allow collaborative work and the joint construction of knowledge. Platforms such as ZOOM and MEET allow constant interaction between the teacher and students and between students synchronously, as well as enabling fluid communication between the educational community. What is still important are the tools and new technologies, which allow the design of teaching materials, the development of synchronous classes and interaction on educational platforms through small devices such as mobile phones, allowing people who do not have access to a tablet or a mobile phone to use them. computer can achieve new learning from his or her phone. The platform are instrument of new computer and communication technologies on the one hand, the systematisation of all processes in an innovative scheme of quality control through language evaluation at each step of vocational training on the other hand.

The contribution of technology in mathematics learning .

Technology helps students learn mathematics, for example, calculators and computers allow students to look up more examples and representations of shapes than by hand, making it easier for them to explore and deduce. The graphical capabilities of technological tools provide access to powerful visual models that many students are unable or unwilling to create on their own. The ability of technical tools to perform calculations broadens the range of problems accessible to students, allows them to perform routine procedures quickly and accurately, and frees them from the development of mathematical concepts and models. Technology can facilitate students' level of engagement and appropriation of abstract mathematical ideas. The opportunity to look at mathematical thinking from different perspectives enriches the scope and quality of their research. According to (Ricardo Poveda, 2017) provides us with the following information: Certain technologies, of a certain general application, are once again appearing on the horizon of education. Calculators and computers seem more "controllable" than television. They may look more promising, but they must always be considered with the pressing need for experimentation to find the optimum point of use. Not only this, but it is also necessary to scrutinise which technologies might be more appropriate to our environment. There are several, for there are also overhead projectors, video projectors, televisions with screens larger than 1 metre, liquid crystal display (LCD) flat screen televisions, video projectors, the computer itself in laptop mode, and others. Although they have been coming down in price, they are still not very easy for institutions to acquire. It does not make much sense to buy them all, but one should (or rather can?) choose very wisely the one that is economically and culturally adapted to a specific condition. Nor does it make sense to underestimate the use of technologies in educational environments: if we look at it properly, technologies help in a number of situations, from the simple presentation of a subject, to the use of them by students in an environment suitably programmed by the teacher.

Student learning is facilitated through feedback provided through technology also plays a central role in the discussion among themselves and with teachers about the objects that appear on the screen and the effects of the various dynamic transformations that technology allows teachers to tailor lessons to the specific needs of their students.Students who are easily distracted can improve their concentration by completing tasks on the computer, and students with organisational problems can benefit from the limitations presented by the computer environment. Students who have difficulty with basic procedures can develop and demonstrate other forms of mathematical understanding that will ultimately help them to learn the procedures, and the use of special technology greatly expands the possibilities for teaching mathematics to students with physical disabilities.

MATHEMATICS TEACHING AND LEARNING

Mathematics education plays an important role in cultivating human resources that can meet the scientific and technological demands of today's social development in that sense, students need to learn. On the other hand, the lack of motivation to study mathematics and the slow development of The skills in this area are obstacles to the achievement of these objectives and the difficulties that mathematics teachers have to face systematically in the exercise of their profession.The publication (Condori, 2021) indicates that: The teaching and learning of mathematics has always had a pre-eminent place in school, although traditionally it has not been the most popular discipline among students, it has been perceived as a less useful knowledge in everyday life, it is the one with the most failures in almost all countries, moreover, it is judged as a subject accessible only to advantaged students, so much so that in some cases it has been used as a measure of students' intelligence (Adamuz and Bracho, 2014; Martínez, 2010). Precisely because of these difficulties, the teaching of mathematics at different levels has been and continues to be a source of concern for institutions, parents and teachers.This definition takes into account the need for students to develop knowledge to make sense of what they are learning if we try to teach students to project and solidify the reality around them, without looking for analogies with the real world, without appreciating the concepts of points and lines that students intuitively elaborate, what we may achieve is that students learn by repetition, but eventually become incapable of responding to the problems they are learning. are presented to them for the first time.Well, if it is true that meaningful learning is effective at the levels identified by the researchers, its possible implementation in higher education should not be ruled out; teachers, unlike students at previous levels, develop strategies aimed at responding to the demands and motivations that instil the use of technology in education. Whether the educational content is considered suitable for linking to real-life situations or other fields, careers followed by the student or historical issues related to learning mathematics, of course, the question of how to apply meaningful learning arises; by identifying students' prior knowledge relevant to what they want to learn, checking whether students master this knowledge and deploying activities if reactivation is difficult, planning differentiated activities for students with difficulties, considering that this has to be related to practice and other fields, or to the historical development of mathematics itself, and cannot be solved with existing knowledge.The use of computer tools motivates students from the start and enables differentiated instruction in mathematics teaching, thus favouring meaningful learning. We have the great challenge of defining methods and strategies that allow us to provide all the resources we need. Computer Science and ICT (Information and Communication Technologies) advocate the integration of the ICT trinomial of students and teachers for meaningful learning in education. Mathematics.

Effective mathematics teaching.

Assessment of learning has come to impose certain artificial responsibilities on the learner in the course of education, irrelevant to the principles and aims of pedagogy, especially of mathematics education, interest in self-learning mathematics has diminished significantly meaning that the responsibility for learning mathematics, and in many cases for learning in general, tends to be greatly reduced. Flexibility in mathematics education should not be limited to these two cases; it is also important to consider the evolution of the problem and its investigation, whether the solution is correct or partially correct, through educational flexibility also includes praise and recognition of student participation and creative solution strategies. According to information from (MORA, 2003), it says that: the preparation of theThe teaching units in the field of mathematics require adequate didactic and special knowledge of the disciplines that could be involved in the intra- or extra-mathematical problems and situations. The solution of such problems must always be understood within the framework of the corresponding mathematical knowledge, which facilitates learning considerably, without causing frustration or didactic rejection. This does not mean that we cannot resort to general solutions and previously established models, which facilitates the solution of the problems generated by the corresponding subject matter.It should also be taken into account that each new situation leads to obviously unexpected or unknown solutions. It is the teacher's task to foresee, to a certain extent, the didactic events that may occur during the development of learning and teaching activities. In this respect, teachers require not only disciplinary, didactic and pedagogical preparation and knowledge, but above all sufficient time and didactic resources. Today, thanks to various studies carried out in the field of mathematics education, many girls and young women do indeed have mathematical difficulties, sometimes very pronounced, despite the importance of comprehensive mathematics education. For the subject and society in general. However, these can be tackled through the development of didactic work in the classroom using collective and individual teaching and learning methods, always adapted to the differences and specific characteristics of the group. On the other hand, it is also important to note that it is not only students who need support who face great difficulties, but also those with a strong interest in mathematicsBoth students and teachers have a decisive impact on the success of the mathematics teaching and learning process; both are responsible for the development and outcomes of the educational practice; they have to accept their strengths and weaknesses and respect each other in their work, learning and teaching. Responsibility for one's own learning and free education does not imply the existence and acceptance of didactic inabilities, on the contrary, it will require more attention from students and teachers. Critical and progressive teaching requires more action in the process and a better sense of content, especially mathematical content. The difficulty of learning mathematics is largely related to the low level of activity in students' mathematical activities, so this is a didactic problem that can be solved by progressive concepts in pedagogy.

THE INTEGRATION OF ICT IN MATHEMATICS

The purpose of primary and secondary education should be for students to acquire the "mathematical skills" necessary to understand, use, apply and communicate mathematical concepts and procedures. The ability to achieve, through exploration, abstraction, classification, measurement and inference, results that enable communication, interpretation and expression. It means discovering that mathematics is relevant to life and the circumstances surrounding it outside the school walls.We need to push for a change in the way mathematics is taught. Teachers need to consider best practices for teaching mathematics, as traditional mathematics teaching has proven to be ineffective. advanced tools. A rich community of mathematical resources. Design and construction tools are also tools for exploring complexity, most of which are freely available on the web. The integration of Information and Communication Technologies (ICT) in the science of Mathematics, this subject, in comparison to the field of action of Language, is fundamental in the immaterial expansion of the students. offer tools to 'become enlightened to think' and to 'become enlightened to become enlightened. Among the curriculum subjects, mathematics has traditionally been a standard-bearer complaint for educators, parents and students, a pervasive measure of students feel shock and poverty of affection when confronted with this subject.The tests applied to the students show that there is a lot to be done to achieve better results in mathematics.

It is evident that the students comfortably carry out simple operations involving one or more variables, they already have problems when they have to connect complex variables and they have to read, add and modify graphics in the power of problems.Under this scheme indicates (Jimenez, 2018) that: Although ICTs are of great social and economic importance, they are not adequately and satisfactorily used in the United States and many other countries around the world. Computer technology is part of the environment in which life takes place, so it is increasingly urgent to learn to live with it and exploit its undoubted potential. The different heights of instruction are intended to enable students to achieve the mathematical competencies necessary to understand, to use, attribute and inform mathematical concepts and procedures that can through exploration, abstraction, classification, proportion and estimation, concentrate on results that allow them to be sincere and execute interpretations and representations; it is to propose, to hit the nail on the head that mathematics if related to the semblance and to the situations that surround them, it is to propose making their training meaningful.

CONCLUSIONS AND RECOMMENDATIONS

At the end of the following research work, the following conclusions and recommendations on the impact of technology as a pedagogical tool to facilitate the learning of mathematics can be drawn.

Conclusions

All pedagogical technological support must have a clear and precise objective in order to meet their expected expectations, otherwise feedback will be the key and fundamental piece to reach students with the knowledge they need. Teachers need continuous training under pedagogical and methodological concepts that allow them to respond to the training of human beings. with integrity, with sufficient capacities to insert themselves productively into society.The subject of mathematics allows the interpretation of reality, this is achieved when the link of dependencies is transformed into a link where everyone contributes the best of their own, normally the institution gives students the commitment to their learning and to a particular discipline within the educational community.

Recommendations.

Avoid comparisons of their abilities with those of other students when reviewing their grades, help them to confront and reward their efforts to understand basic concepts through words of encouragement and congratulations. Parents should solve with a practice of exercises and actively participate in the training of their children, on the web there are mathematics lesson programmes addressed to parents, where recommendations for correcting in mathematics in primary school are offered. Teachers have the task on a scale of methods, techniques and educational strategies that allow them to handle the growth of the students' numerical rational argument and to teach them to love the seriousness of the examination of this subject, as well as to take responsibility with simple strategies in the pedagogical praxis to activate the lesson, considering the new educational realities, the characteristics of their students and the environments where they develop. The benefit of playful strategies in the first grades of basic training, in order to make the preparation of mathematics fun, activities such as colouring figures to confront geometry, educational game days can be planned for the students to practice, solving problems that stimulate the growth of mental speed.

BIBLIOGRAPHY

Acuña, B. P. (January 2011). Scientific methods of observation in education. Retrieved from https://www.researchgate.net/.../262817848_Metodos_cientificos_de_obs ervacion_en_E...

Alícia Marti (6 October 2020). How to incorporate ICT in the classroom: concrete proposals easy to apply. Retrieved from https://docentes.algareditorial.com/blog/59/incorporar-tic-aula- propuestas-concretas-faciles-docentes

Ana VIÑALS BLANCO, J. C. (2016). The role of the teacher in the digital era. Retrieved from https://www.redalyc.org/journal/274/27447325008/html/

Angulo, J. S. (13 January 2017). On population and sample in empirical research. Retrieved from https://cuedespyd.hypotheses.org/2353

Arnoletto, E. J. (2013). Los conflictos en los procesos sociales. Cordoba Argentina: Fundacion Universitaria andaluza Inca GArcilaso for Eumed.net.

Berchon, M. (2014). The 3d printing. Editorial Gustavo Gili, SL, Barcelona, 2016.

Chagoya, E. R. (2008). Diseño teorico metodologico de la investigacion. Retrieved from Gestiopolis: https://www.gestiopolis.com/diseno-teorico-metodologico- de-la-investigacion/

Christ, A. J. (2007). Culture of peace and educational reforms. Retrieved from www.uasb.edu.ec/.../educacionenyparalosderechoshumanos/.../Culturadep azyreformas....

Companies, S. d. (December 2012). Measurement guide. Retrieved from http://appscvs.supercias.gob.ec/guiasUsuarios/images/guias/proc_cm/guia

_mediacion.pdf.

Condori, A. P. (2021). Learning Mathematics for the 21st century. Retrieved from http://repositorio.unae.edu.ec/bitstream/56000/2122/1/Didacticasmatematicas-17-48.pdf

Correa, A. G. (2004). La mediación: Técnica de resolución de conflictos en contextos escolares. Anuaria de Filosofia, Psicologia y Sociologia, 10.

EcuRed (11 May 2018). Poblacion. Retrieved from https://www.ecured.cu/Poblaci%C3%B3n

Eduardo Parra Zambrano, R. P. (2015). Curricular integration of ICT. Retrieved from https://www.oas.org/cotep/GetAttach.aspx?lang=es&cId=412&aid=707

Education, M. d. (28 September 2012). www.educacion.gob.ec. Retrieved from https://educacion.gob.ec/wp-content/plugins/download- monitor/download.php?.

Education, P. (21 July 2020). Teaching and the future of programming. Retrieved from https://pinion.education/es/blog/ensenanza-y-futuro-de-la-programacion/#footermovil

Graterol, R. (2011). La investigacion de Campo. Merida State of Merida Venezuela.

Gutierrez, A. C. (2011). La mediacion internacional en el sistema de las naciones unidad y en la Unjon europea:Evolucion y retos de futuro. Revista de mediacion, 29. Retrieved from www.ammediadores.es/nueva/wp-content/uploads/2014/12/Revista-8.pdf

Instroduction: Children in the digital age (2017). Retrieved from Unicef: chrome-extension://efaidnbmnnnibpcajpcglclefindmkaj/https://www.unicef.org/media/48611/file

Jimenez, J. J. (2018). Las tics: a new challenge for the classroom. Retrieved from chrome-extension://efaidnbmnnnibpcajpcglclefindmkaj/https://archivos.csif.es/archivos/andalucia/ensenanza/revistas/csicsif/revista/pdf/Numero_13/JUAN_J_BAENA_1.pdf

Joranporre (2015). Investigacion Bibliografica. Retrieved from http://mtu-pnp.blogspot.com/2013/07/la-investigacion-bibliografica.html

Lopez, P. L. (2004). Scielo Bolivia. Retrieved from Poblacion Muestra y Muestreo: www.scielo.org.bo/scielo.php?script=sci_arttext&pid=S1815....

Basic Mediator Training Manual (18 May 2009). Retrieved from https://www.santafe.gov.ar/index.php/web/content/download/71289/345 896

Maricel Occelli, L. G., & Gatica, M. Q. (2019). INFORMATION AND COMMUNICATION TECHNOLOGIES as mediating tools for educational processes. Retrieved from http://biblioteca.clacso.edu.ar/clacso/se/20191128031455/Tecnologias- digital.pdf

Martin, F. a. (2011). The survey: a general methodological perspective. Spain: Caslon S.L. Matilde Hernandez.

Mateo, S. M. (2010). Who we are, where we are going.. origin and evolution of the
concept of mediation. Mediation, 8.

Merino, P. y. (2010). Research. In P. y. Merino.

Monterrey, T. (2005). Strategies for effective mathematics teaching. Retrieved from http://www.cca.org.mx/ps/profesores/cursos/depeem/html/contenidos/module1/theme1-3.htm

MORA, C. D. (2003). Strategies for learning and teaching mathematics. Retrieved from Revista Iberoamericana de Educación Matemática: http://ve.scielo.org/scielo.php?script=sci_arttext&pid=S0798-97922003000200002

NUEZ, M. B. (2008). OEI - Revista Iberoamericana de Educación. Retrieved from "ESTRATEGIAS EDUCATIVAS PARA EL USO DE LAS NUEVAS TECNOLOGIAS DE LA: https://rieoei.org/RIE/article/download/3008/3911/

Paula, C. G., & Caldeiro, P. G. (2014). SCHOOL MEDIATION. Retrieved from https://educacion.idoneos.com/355341/

Peña, L. B. (2010). The literature review. Colombia.

Projects, F. y. (13 June 2011). Feasible Project. Retrieved from http://proyectofactible6.blogspot.com/

Ricardo Poveda, M. M. (2017). NEW TECHNOLOGIES IN THE TEACHING AND LEARNING OF MATHEMATICS. Retrieved from https://www.centroedumatematica.com/aruiz/libros/Uniciencia/Articulos/Volume1/Part6/article10.html

Rossana, C. A. (January 2021). Atlante Journal: Notebooks of Education and Development . Retrieved from El uso de las Tic en el proceso de Enseñanza y Educación.

Learning: chrome-extension://efaidnbmnnnibpcajpcglclefindmkaj/https://www.eumed.net/uploads/articles/24f38807a68414015be264023a0fb0b9.pdf

Ruiz, P. (30 October 2012). Retrieved from Que es la tecnologia: https://aptandalucia.wordpress.com/2012/10/30/que-es-la-tecnologia/

Seminarium, F. E. (2018). Digital tools: The new challenges and opportunities in 21st century education". Retrieved from https://www.fundacionseminarium.com/herramientas-digitales-los-nuevos-desafios-y-oportunidades-en-la-educacion-del-siglo-xxi/

Tamayo, M. T. (2003). The process of scientific research. Balderas, Mexico: Limusa S.A. Grupo Noriega Editores.

Torrengo, J. (2001). Mediation of conflicts in educational institutions. Madrid: Narea S.A.

Unesco (2014). United Nations Educational, Scientific and Cultural Organization. Retrieved from Theoretical and Practical Module on School Violence Prevention and Conflict Resolution: http://unesdoc.unesco.org/images/0024/002471/247141s.pdf.

Vega, J. (13 June 2018). Pedagogical tools, a resource to enhance children's development through play. Retrieved from https://maguared.gov.co/las-herramientas-pedagogicas-un-recurso-para- potenciar-el-desarrollo-el-los-ninos-por-medio-medio-del-juego/

VIVES, J. (23 June 2021). Robotics as an educational tool. Retrieved from LA Vanguardia: https://www.lavanguardia.com/vida/junior-report/20210623/7551118/robotica-herramienta-educativa.html

Printed by Books on Demand GmbH, Norderstedt / Germany